Noivado Beula
Anussa Mirasse
Naftal Zefanias

The environmental impact of opencast coal mining

Noivado Beula
Anussa Mirasse
Naftal Zefanias

The environmental impact of opencast coal mining

Case study: Moatize District (2010 to 2023)

ScienciaScripts

Imprint
Any brand names and product names mentioned in this book are subject to trademark, brand or patent protection and are trademarks or registered trademarks of their respective holders. The use of brand names, product names, common names, trade names, product descriptions etc. even without a particular marking in this work is in no way to be construed to mean that such names may be regarded as unrestricted in respect of trademark and brand protection legislation and could thus be used by anyone.

Cover image: www.ingimage.com

This book is a translation from the original published under ISBN 978-620-6-76037-5.

Publisher:
Sciencia Scripts
is a trademark of
Dodo Books Indian Ocean Ltd. and OmniScriptum S.R.L publishing group

120 High Road, East Finchley, London, N2 9ED, United Kingdom
Str. Armeneasca 28/1, office 1, Chisinau MD-2012, Republic of Moldova, Europe
Printed at: see last page
ISBN: 978-620-7-71736-1

Table of Contents

Summary

This article is about the environmental impact of opencast coal mining. Its general objective is to assess the environmental impacts caused by opencast coal mining in the Moatize district. Its problem is what is the environmental impact of opencast coal mining in the Moatize district? We opted for a qualitative approach in an exploratory and descriptive study, and the technical procedures consisted of a case study. It was concluded that living in Moatize is unsafe due to the harmful effects that coal mining can have on them. Dust was a very common element among the respondents. The results of the study show that it is imperative that there is an increase in the institutional responsibility of companies and government bodies, providing health care and promoting environmental interventions for the population of Moatize, directly affected by the precarious living and health conditions and by coal dust, as well as a more in-depth re-discussion of the political and ethical aspects linked to the resettlement of the most exposed population groups.

Keywords; Mineral coal, exploitation, Environment and open-cast.

Introduction

Contextualisation

According to Goldemberg (1998), we can see that the evolution of humanity, from the time of primitive man to contemporary society, is directly linked to energy consumption. We can therefore say that energy is fundamental to the development of life.

In this way, the introduction of mineral coal marked the end of the era of renewable energy represented by wood and the meagre use of hydroelectric and wind power, and the beginning of the non-renewable era of energy - the era of fossil fuels (Bellia, 1996).

Mineral coal is therefore one of the most important resources for a nation's development, but obtaining it jeopardises the very characteristics of the nation's environment and nature.

Nature has always had forms of self-regeneration that conform to the natural dynamics of the transformation of natural resources. However, this capacity has been surpassed by human activities which, given the growing need to increase the efficiency of natural resource exploitation and to overcome the limitations of these resources in the face of current demographic growth and, above all, economic and financial imperatives, have introduced variables that are foreign to natural dynamics. These changes have contributed greatly to the deterioration of the natural environment.

However, for the rational use of coal, technical and scientific measures have been developed that allow for sustainable exploitation, the correct interpretation of natural phenomena and their optimisation.

Coal mining in Mozambique has increased rapidly in the last 10 years, due to the discovery of new coal mine deposits in Tete province. In 2008, it was discovered that one of the largest untapped coal basins in the world, with an estimated capacity of around 23 billion tonnes of coal, is in the Moatize district of Tete province.

In 2010, multinational companies such as Vale Moçambique (from Brazil), Rio Tinto (Anglo-Australian), Jindal Steel and Power (Indian), Coal and Beacon Hill (UK), Eurasian Natural Resources Corporation (Kazakhstan) and Minas de O Revubué (Anglo-American) established themselves in Tete Province to exploit coal (Pondja, 2017).

Nature always has forms of self-regeneration that conform to the natural dynamics of the transformation of natural resources. However, within the Mozambican reality, this capacity has been surpassed by human activities which, given the growing need to increase the efficiency of the exploitation of natural resources and to overcome the limitations of these resources in the face of current demographic growth and mainly due to economic and financial imperatives, have been introducing impacts on natural dynamics. These changes have contributed greatly to the deterioration of the natural environment.

The choice of topic was justified by the authors following frequent complaints from residents in the Moatize region regarding open-cast coal mining activities by Vulcan and ICVL, which are related to the quality and non-compliance seen in the process of changes in ecosystem services, in particular changes in water and air quality.

The study of the subject is relevant not only for the professional and personal development of the researchers, but it is also important for the social context, since every open-cast coal mining activity has an impact on citizens, thus becoming a mechanism for informing the public interest that seeks to improve knowledge about open-cast coal mining activities.

In 2017, the Moatize Coal Mine produced around 11 million tonnes of coal. As opencast coal mines, their environmental impacts include air pollution, water pollution, soil degradation, social impacts and carbon emissions that contribute to climate change. On the other hand, communities complain about the negative impacts caused by opencast coal mining in Moatize.

This is the purpose of a case study to help understand this problem.

In view of the above arguments, the need arose to answer the following questions: What is the environmental impact of opencast coal mining in the Moatize district? To provide further grounding, we also had the following questions: What are the potential ecosystem services associated with the area covered by the project? Are mitigation actions undertaken by both companies within the framework of their respective Environmental Management Plans?

In order to answer this question, the following general objective was set: To assess the environmental impacts caused by opencast coal mining in the Moatize district, which is sub-categorised into the following specific objectives of the study: To identify the potential ecosystem services associated with the project area; To describe how the effects of coal mining interfere with the supply of the ecosystem services identified and the well-being of the local population; To understand the mitigation actions undertaken by both companies, within the framework of their respective Environmental Management Plans.

Apart from this brief introduction, this article is organised into five sections. The second section presents the literature review, including empirical studies. The third describes the methodological considerations.

The fourth section shows the results and discusses the study and, finally, the fifth section presents the final considerations.

Theoretical Foundations

A brief introduction

In this literature review chapter, we will first discuss the concepts of the environment, mining, opencast mining, coal mining and environmental impact, according to various authors, followed by their theoretical foundations.

Presentation of basic concepts

Environmental impact

Coelho (2004) defines environmental impact as "any alteration to the physical, chemical and biological properties of the environment caused by any form of matter or energy resulting from human activities that directly or indirectly affects: the health, safety and well-being of the population.

Thus, we understand environmental impact as any alteration produced by humans and their activities in the relationships that make up the environment, which exceed the environment's capacity to absorb.

Environment

For Marques (2005), the environment is the sum total of the surrounding external conditions within which an organism, a condition, a community or an object exists, adding that organisms can be part of the environment of other organisms.

According to Mendonça (2004), the environment no longer receives the traditional descriptive or contemplative view from geography as if it were a sanctuary that exists parallel to society.

Therefore, the environment, since it includes man and everything that surrounds him, is a dynamic and ever-changing process, caused both by external factors, without the influence of man, flora or fauna, and by the actions of human beings in the transformational processes of the raw materials they manipulate, as well as cultural transformations caused by changes in values induced by man himself.

Mining

The mining sector is very diverse, both in the form in which resources are found in nature and in their volumes and characteristics, so the description of the extraction process must be specific to each substance. According to Abrão and Oliveira (1998), mining is understood as an activity involving the extraction and concentration of minerals.

Mining is an extractive economic activity characterised by the exploitation of natural resources, supplying humanity with mineral inputs in the form of raw materials, energy, fillers, noculants and other accessory substances in a wide variety of industrial and construction processes, which are fundamental to the quality of life and socio-economic development of modern society (Girodo, 2005).

Mining is the process of extracting minerals that are naturally concentrated in the earth. It is an economic and industrial activity that consists of research, exploration, ploughing (extraction) and processing of minerals found underground (Calaes, 2006).

As such, mining is a term that covers the processes, activities and industries whose aim is to extract mineral substances from proven mineral deposits in the ground or subsoil.

Open-pit mining

Open-pit mining is nothing more than extensive excavation of the land surface in order to extract metallic and non-metallic minerals from any type of rock. Open-cast mining can range from small manual scrapes on the surface of the ground to gigantic excavations that reach hundreds of metres in depth and can occupy tens or even hundreds of square kilometres on the surface (Girodo, 2005).

Open-pit mining is work carried out by men with shovels and picks or by large mining equipment (Redaelli & Cerello, 1998).

Mineral coal

Coal is a dark-coloured sedimentary rock formed by a complex and varied mixture of organic components fossilised over millions of years, characteristic of fossil fuels. These components are the remains of prehistoric vegetation that accumulated in swamps or muddy regions, under a sheet of water. Over time, these deposits were covered by clays and sands (generating sedimentary basins), this gradual coverage led to an increase in temperature and pressure that expelled oxygen and hydrogen from the environment, concentrating the carbon, this process called carbonisation (Schumann, 2008).

Mineral coal is composed of oxygen, hydrogen, sulphur, ash and, for the most part, carbon. The main constituent of mineral coal is wood, which is responsible for the percentage of carbon in its composition. Wood is made up of approximately 50 per cent carbon, 44 per cent oxygen, 5 per cent hydrogen and 1 per cent other compounds (Schumann, 2008).

According to Viana (2005), mineral coal can be conceptualised as a combustible sedimentary rock originating from organic matter. The processes that convert this organic matter into coal are called carbonification processes and require variables such as pressure, temperature, time and the action of micro-organisms.

Coal can be defined as a combustible sedimentary rock made up of certain plants, as well as woody tissues and hydrocarbons, i.e. heterogeneous material that has been buried and compacted in originally shallow basins. Factors such as pressure, temperature and tectonics can determine the gradual carbonisation of the original material, which has undergone changes such as loss of oxygen, water and carbon enrichment.

A mineral can be defined as an inorganic crystalline solid that occurs in nature, with a well-defined chemical composition and characteristic physical properties (Wicander & Monroe, 2009). According to Schumann (2008) a crystal is a materially uniform body with a regular structure of its minimal particles (atoms, ions or molecules).

According to Braga, França and Sampaio, (2010) Ore Treatment or Beneficiation consists of operations applied to mineral goods with the aim of modifying the granulometry, the relative concentration of the mineral species present or the shape, without however modifying the chemical or physical identity of the minerals.

The current situation in the area of environmental control of emissions in the mining and metallurgical sector is characterised by the existence of research, procedures, technological processes and human resources that make it possible to meet current legislation in a satisfactory manner. Efficiently treating, recycling or reusing the effluents generated has become essential for the industrial sector, which is seeking to combine good acceptance of its products on the international market with compliance with current environmental legislation. In the case of liquid effluents, there is also a reduction in operating costs due to the future charging for the use of water at source. However, the industrial sector has yet to absorb knowledge of new technologies and often insists on applying conventional, low-efficiency methods.

According to Braga, França and Sampaio (2007), the majority of mineral goods used by society have low added value; therefore, the use of technology and low-cost inputs become indispensable requirements to guarantee the economic viability of their exploitation. According to this approach, water is the most important input in mining, especially in the area of ore concentration, where it is basically used as a means of transport.

According to the aforementioned authors, there are several factors that justify this application, including its still high availability and low cost of collection. Of equal interest are its chemical and physical characteristics, including: density, viscosity, chemical composition, conductivity, vapour pressure, surface tension, freezing and boiling points. Although it is used less than in agriculture, water is essential for most mineral processing operations. Even though it may seem elementary, this is a requirement that cannot be overlooked by mining professionals. Modern processing plants are increasingly demanding better quality water in water/ore ratios ranging from 0.4 to 20. The amount of water needed for many ore processing operations, as well as its quality, are determined by studies on laboratory and pilot scales.

According to Braga, França and Sampaio (2007), water is the most commonly used liquid phase in ore processing plants that use wet separation processes (gravitational, magnetic, flotation, flocculation, spherical agglomeration, leaching and others).

Its availability is one of the basic requirements for mineral processing, as well as being a determining factor in the location of the mineral processing plant. Therefore, a reliable and adequate supply of water and its storage and transport become indispensable in mineral processing

Operations that can cause environmental impacts in mining and mineral processing engineering

According to Braga, França and Sampaio, (2010) in the activities of Mineral Engineering and Extractive Metallurgy some operations can cause environmental impacts:

- ✓ Prospecting and exploration;
- ✓ Mine development and ore extraction;
- ✓ Waste rock disposal, decommissioning/reclamation of areas;
- ✓ Fragmentation or comminution: crushing and grinding;
- ✓ Size control: classification;
- ✓ Physical/physico-chemical concentration;
- ✓ Unit operations associated with the treatment of concentrates: thickening, filtering, drying, agglomeration and tailings disposal;
- ✓ Hydrometallurgy, pyrometallurgy, biometallurgy, electrometallurgy and combined processes.

Environmental impacts of mining operations

According to Braga, França and Sampaio (2010), the main impacts of mining operations are noise, dust and, in some cases, soil contamination by acid drainage and explosives derived from glycerine, which produce heavy metal ions such as copper, nickel, lead, zinc, mercury, as well as iron and anions such as sulphate, phosphate, arsenate, tellurate, fluoride, molybdate, cyanide, among others. Also, oil, grease and organic solvent spills, gas emissions, plastic waste, sludge (precipitates), acid-producing tailings, visual pollution, biodiversity changes, tailings disposal-storage (ponds, pits), radioactivity and spontaneous combustion (pyrite from coal, for example) are common impacts from this activity.

2.2.2.2 Environmental impacts of processing operations

According to Braga, França and Sampaio (2010), the most common impacts of beneficiation are caused by liquid effluents containing heavy metals and toxic anions, solids (sometimes colloidal), organic waste (foaming agents, collectors, surfactants and oils). We can also highlight the production of dust and noise in the comminution stages, gaseous emissions (organic) and vapours, such as mercury, which even in small proportions has a very significant impact.

Environmental impacts in Extractive Metallurgy/Metallurgical Processing operations

According to Braga, França and Sampaio (2010), environmental impacts in metallurgy come from liquid effluents containing heavy metals, anions, organic and metallic compounds, toxic complexes and chelates, emulsified or non-emulsified oils (organic solvents), fine or colloidal solids (sludge), as well as gases (SO_2, NOx) and aerosols (arsenic oxides, volatile ashes) and the deposition of solid waste (pyrites).

Methodological Procedures

Freitas and Prodanov (2013), applied research aims to generate knowledge for practical application aimed at solving specific problems. It involves local truths and interests". The research followed a qualitative approach. In terms of objectives, the research is exploratory and descriptive. Descriptive research is seen as "research aimed at finding the existence of relationships between variables (Gil, 1994). In terms of procedures, this is a case study and a bibliographical survey.

For this study, the residents of the Moatize district were involved as the research universe. The inclusion criterion was that they were direct stakeholders in the internal communication process.

However, for this research, which focused on collecting information from twenty-two (22) participants (residents) living in the district of Moatize, we used non-probabilistic sampling, but rather simple convenience sampling.

For this research, the data collection techniques used were the questionnaire and observation. For this study, it was necessary to use the questionnaire survey, as it is one of the most reliable techniques for this type of research. All the questionnaires were semi-structured with mixed questions characterised by open, closed and multiple choice questions. The respondents to this questionnaire were the residents and, thanks to the simplicity of the author, they were free to express and justify their opinions on the subject freely when answering the open questions in the questionnaire. The questionnaire was addressed to the head of the neighbourhood and the residents of the Moatize district.

The study is the result of extensive direct observation. This technique allowed the researcher to get closer to the subject's perspective, to follow the daily experiences of the individuals included in the research. According to Lakatos & Marconi (1991), extensive direct observation uses the senses to obtain certain realities. It doesn't just consist of seeing and hearing, but also examining the facts and phenomena that one wishes to study.

In this case, through observation, the author experienced the reality of the Moatize district, trying to identify potentially affected ecosystem services and find elements that support the environmental impact of opencast coal mining in the Moatize district.

And through this, the researcher, without needing to contact the stakeholders, checked whether or not they fulfil their obligations in terms of reducing environmental impact and improving the quality of ecosystem services.

In this work, the data was processed and analysed using descriptive content analysis, as Gil (2002) states that techniques "are operational procedures that serve as practical measurement or data collection and analysis for carrying out research (p. 26).

According to Bardin (2011), this technique is very didactic and facilitates the sequence of tasks and activities to be followed in order to analyse qualitative data (p. 15).

Presentation, analysis and discussion of the results of the case study.

This chapter presents and analyses the research data in a qualitative way. As a methodology to make it easier to compare the results, the tables and respective graphs are presented.

Perception of environmental risks

The environmental risks were identified on the basis of observations in the field and responses about the changes that have taken place over the last five years in the neighbourhood's environment. Thus, the main environmental risk mentioned by respondents was air contamination from dust.

Presentation of environmental quality data

Of the 22 respondents, only 3% said that there had been no environmental changes in the Municipality of Moatize. The majority of respondents (97%, 43% women and 54% men) said that there had been environmental changes in the neighbourhood where they live and that there had been improvements in some aspects of the neighbourhood's environment: 94% said that the improvements were due to good water quality, (3%) soil quality and (3%) less dirt.

Regarding changes in the neighbourhood's environment for the worse, 93% of respondents indicated that air quality had worsened; (4%) soil quality.

The analysis of the score, on a scale of 0 to 20 points, attributed to the neighbourhood's environment indicates an average of 8.24 points and ranks third in the score of the worst changes that have occurred in the Municipality of Moatize. However, 59% of those surveyed said that the neighbourhood environment was good, 12% said it was bad and 18% said it was worse because of the dust.

Presentation of water quality data

With regard to the environmental component water, the average score was 16.83 points and it is the first environmental aspect that has improved, according to 94% of the respondents (43% of women and 51% of men), in the last five years in the Municipality of Moatize. However, some of the women surveyed (3%) were concerned that they were facing difficulties in accessing water.

In general, respondents don't seem to have a perception of the impact of coal dust on water quality, since their assessment is linked to access to and supply of water and not to water quality per se.

Respondents reported that the improvement in the water quality component is due to having water every day (30 per cent), that it is better (60 per cent) and that it is normal (3 per cent). Despite this, the vast majority of respondents reported that the quality of the water is good because it is treated and even those in rural areas seem to be satisfied with the water deposited in their reservoirs.

Presentation of soil quality data

With regard to the quality of the soil, the majority (64 per cent) of respondents replied that the land did not allow for good agricultural production, some (15 per cent) said it was good, while others indicated that the soil was impermeable. What's more, 9 per cent of respondents reported that their low score was due to the fact that there was "rubbish".

Therefore, the respondents' perception of soil quality does not establish a direct relationship with the contamination of the soil by coal dust. However, the soil quality score ranks second as one of the aspects that has worsened the most in the municipality of Moatize over the last five years, after air quality.

Presentation of air quality data

In the air quality component, the average score was 3.82 points, the lowest of the evaluations. According to 83% of those surveyed (42% women and 51% men), air was the environmental element that stood out as the worst environmental change in the municipality of Moatize.

The change for the worse is due to the pollution caused by coal dust from the extraction of mineral coal in Vale Moçambique's developments. The perception of air quality varied from extreme positions, from clean air (3 per cent of respondents) to polluted air (53 per cent of respondents). Respondents said that the air was good (3%), that there is no good air in Moatize and that the air has coal in it (6%), that the air causes coughing (7%), that the air is dusty (53%), and that it causes breathing problems (18%).

Dust was a very common element among those surveyed. It is often referred to as a popular form of coal dust, and causes health problems for those who live in the municipality of Moatize, especially in the neighbourhoods close to the mine.

As MONIZ (2010) cites (Porto and Freitas, 1997; Marandola Jr. and Hogan, 2004; Brilhante and Caldas, 1999), the individual and social perception of environmental impacts and actions to mitigate them (adaptations and adjustments) have losses and gains on different scales, especially when the causes of the risks are industrial technological processes.

Perceptions of health risks

Presentation of coal dust data

Alongside the environmental problems mentioned above, the majority of respondents perceived coal dust as an environmental factor that jeopardises human health and the environment in the Municipality of Moatize. With regard to the negative impacts of coal dust on the environment, 93 per cent of those surveyed had a perception of the negative impacts of coal dust.

Presentation of data on living in Moatize

According to Moniz (2010), citing Oliveira and Zambrone (2006), the territory and the social, environmental and personal factors that contribute to structuring social life have a direct influence on the degree of vulnerability in relation to a given health-related event, as a function of individual and collective control over the risk of acquiring a disease.

Thus, the workplace and place of residence can be contexts for the occurrence of health problems for the individuals who work there. According to this understanding, the results of the questionnaire show that residents of Moatize have established a relationship between the place where they work and/or live and the development of events that put them at risk of health problems.

Respondents were asked whether they thought it was safe to live in the Municipality of Moatize or in the community of Benga. Of the 100 respondents, 81% answered that living in Moatize is not safe because there is a greater chance of getting a disease than elsewhere.

It was observed that a larger number (47%) of the residents surveyed perceive that living in Moatize is unsafe. This may be related to some individuals' life experience of the impacts that coal mining can have on them, their friends or family members, even if the mine is not their place of work, and not to their level of education.

According to Moniz (2010), citing Freitas et al, (2003) and Freitas (2000), other "reasons may have contributed to individuals' judgements about the relationship between risks and local security, among which situations of socio-environmental vulnerabilities that are inseparable from population and institutional vulnerabilities stand out".

Description of the environmental impacts caused by companies

Based on the findings of the laboratory analyses of the tailings piles in modules I and II, during the processing of mineral coal at the company's plant in modules I and II, it was possible to observe the occurrence of practices harmful to the environment and the weakness of some mitigation measures. These impacts are directly linked to the inadequate disposal of mineral coal tailings and associated with the unit operations of coal processing:

- ✓ Air pollution (during granulometric preparation (ROM section) and from CO_2 emissions during spontaneous combustion of the coal tailings);
- ✓ Soil contamination during mineral coal processing at the plant in the coarse circuit, the fine circuit, thickening and tailings piles;
- ✓ Contamination of groundwater; during coal processing at the plant in the coarse circuit, the fine circuit, thickening and tailings piles
- ✓ Destruction of the landscape;
- ✓ Destruction of flora;
- ✓ Noise pollution.

Air pollution

During the processing of mineral coal at the company's plant, based on the data obtained, it was possible to see that air pollution occurs at the following stages:

Air pollution during particle size preparation (ROM section)

In the silo of the primary crusher, when the material coming from the mine is unloaded and fed into the ROM section, ultrafine and microscopic particles are ejected, mixing with the air and polluting the air that is breathed and inhaled by the people who work nearby and the people who live near the company's plant.

Before processing itself, the material coming from the mine is subjected to adjustment in order to reduce the particle size and classify the particles into different size classes. At the company, this comminution and classification process in the pre-scalp is done dry, i.e. without the presence or addition of water, thus causing the ejection of ultrafine particles that mix with the air, thus leaving the air polluted.

This air pollution not only impacts on the quality of the air that is consumed in the Moatize region, but also on the development of the region's flora and fauna.

No mitigation measures had yet been implemented for this environmental impact by the time of the study.

It is possible to notice the large emission of ultrafine particles which, when mixed with the air, pollute it by forming dark-coloured dust. This air

The polluted air travels several kilometres to the communities around the mine, leaving people breathing polluted air.

At the company, the only measure to reduce dust is the spraying of water by tanker lorries. This measure isn't even sufficient because no water is sprayed in the ROM section and there is no device installed to purify the air.

Air pollution from CO_2 emissions from tailings piles

One of the main environmental impacts observed after the processing of mineral coal is the disposal of material that contains ash contents of less than 50%, which is considered to be mineral coal that spontaneously combusts when it comes into contact with oxygen.

Module 1 tailings pile (Wet base%)						
Month of analysis	Samples	Total Humidity	Volatile Material	Fixed Carbon	Ash (%)	Ash (g)

November	1	1,24	23,19	43,78	35,791	258,31
	2	1,11	20,88	44,55	33,45	71,91
	3	0,14	24,68	46,17	29,01	150,01
December	1	1,02	18,71	19,95	60,32	109,41
	2	1,2	16,827	29,55	52,43	283,23
	3	2,07	22,10	33,82	60	168,78
January	1	0,29	18,50	28,9	52,3	53,53
	2	2,60	22,71	30,33	55,39	111,39
	3	2,35	23,46	34,22	37,98	17,02
Total	9				416,659	1 223,60
Media		1,34	21,2	34,6	46,3	136,0

Table 1: Immediate analyses of Module 1 tailings

Source: Author, 2023.

Based on Table 2, it is possible to see that the samples of tailings from Module 1 that were analysed have an average ash content of 46.3% below the standard established and considered to be tailings, in this order of ideas, the data shows that the tailings pile in Module 1 contains mineral coal, which culminates in the emission of CO_2 and SO2 because mineral coal, when exposed for a certain period of time, reacts with oxygen and ends up undergoing slow, spontaneous combustion.

Up until the analysis period, the company still had no measures in place to mitigate this environmental impact.

Module 2 tailings pile (wet basis%)						
Month of analysis	Samples	Total Humidity	Volatile Material	Fixed Carbon	Ash (%)	Ash (g)
November	1	0,19	21,19	42,78	50,84	424,42
	2	1,10	19,88	33,46	45,55	179,00
	3	0,13	18,68	51,95	29,24	207,55
December	1	1,02	18,71	30,95	49,32	89,46
	2	0,2	16,827	18,55	64,43	206,62
	3	2,83	22,28	33,82	39,08	109,94
January	1	0,29	18,50	28,9	53,3	27,91
	2	0,92	22,71	26,33	60,34	196,79
	3	1,63	22,46	18,22	37,70	54,59
Total	9					1
Medias		0,92	20,1	31,7	47,8	166,3

Table 2: Immediate analyses of Module 2 tailings

Source: Author, 2023

Table 3 shows that the samples of tailings from Module 2 that were analysed have an average ash content of 47.8% below the standard established and considered to be tailings. In this order of ideas, the data shows that the tailings pile in Module 1 contains mineral coal, which culminates in the emission of CO_2 and SO2 because mineral coal, when exposed for a certain period of time, reacts with oxygen and ends up burning slowly and spontaneously.

At some points of the company's tailings dam, smoke can be seen emitting from the dam containing carbon dioxide (CO_2) and sulphur dioxide (SO_2). When these substances are found in high concentrations in the air, they are super toxic and harmful to the life of any living being. In addition to putting the lives of people in the surrounding areas at risk, spontaneous combustion contributes significantly to the destruction of the ozone layer, thus putting the ecosystem out of balance.

As can be seen in the picture above, smoke is visible, this smoke is the result of the spontaneous combustion of the mineral coal tailings as they contain certain amounts of coal that should have been recovered, i.e. based on the immediate analyses made of the company's tailings, the ash content of the tailings is low, which shows the presence of coal in the tailings, this coal when in contact with oxygen undergoes spontaneous combustion allowing the emission of CO_2 which is one of the biggest agents destroying the ozone layer.

Soil contamination during coal processing at the company's plant in the coarse circuit, fine circuit, thickening and tailings piles

After comminution and classification of the mineral coal, the material is subjected to the actual classification process, with the aim of classifying the particles into species with dynamic physical and chemical characteristics. In this classification process, all the circuits operate with the feed in the form of a slurry (Water + Solid), in the coarse circuit, in order to separate the coal from the tailings, it is essential to create a dense medium in order to separate the particles into different density classes, thus using a dense medium which is a combination of water and magnetite, and in the thickening process a combination of water and some depressors and collectors, in the two modules of the Jindal plant, during processing there have been several leaks in the pipes and even in the equipment itself, which allow chemical substances mixed in the water to be emitted into the soil. These liquids containing these substances accumulate to the point of forming puddles and thus reacting with the soil and thus altering the physical and chemical characteristics of the soils.

The infiltration of these fluids containing these chemical elements ends up chemically reacting with the chemical elements present in the soil and thus causing a change in the physical and chemical characteristics of the soil profile of the area where the water is stagnating.

In the processing plants of the Jindal Africa company, pipes and some equipment leak water mixed with certain chemical substances. At some points in this image, you can see the formation of puddles and standing water that end up reacting with the soil.

Groundwater contamination during coal processing at the Jindal plant in the coarse circuit, fine circuit, thickening and tailings piles.

Almost all of the company's mineral coal processing is done wet, i.e. with the presence of water to make it easier to handle the material and facilitate the process.

In the coarse circuit there are leaks along the pipes that allow certain quantities of pulp to escape, at the end of the dense medium cyclones it was possible to observe the presence of several leaks that allow the dense medium containing water and magnetite to escape, as well as in the dense medium preparation tank there are leaks that allow fluids to escape. These fluids that have leaked along the circuit end up forming accumulations on the floor of the plant which, after a period of time, infiltrate and thus end up contaminating groundwater, because these fluids that migrate to the groundwater contain chemical elements that cause contamination of the groundwater.

One of the major factors culminating in the emission of contaminated water into the groundwater table is the deposition of tailings with a high percentage of humidity, which ends up allowing the infiltration of this fluid that ends up contaminating the groundwater.

Based on research carried out by the company, in some of the boreholes that people consume and use for their daily activities, the water comes out cloudy and darkly coloured, which is the result of groundwater contamination.

This contamination puts the health and lives of consumers of borehole water at risk because it sometimes contains harmful chemicals that are harmful to human life and the animals that consume it.

In the fines circuit, the feed is made in the form of a slurry and with a pressure to allow the material to flow through the spirals. During the analyses carried out, it was clear that in this circuit a large amount of water is wasted and projected onto the ground. These quantities of water end up forming small puddles.

In the company's tailings piles, it is possible to see large quantities of water coming out of the tailings. This is due to the deposition of tailings with high humidity, i.e. the deposition of solids that contain water, and these waters containing chemical substances that are harmful to the environment and human health end up migrating and contaminating the groundwater.

Destruction of the landscape

One of the environmental disasters observed during the company's activities is the destruction of the landscape and topographical features that once existed.

During the preparation of areas for the deposition of tailings from the Jindal Africa plant, the entire vegetation cover is removed and ridges are illuminated in order to level the area for the deposition of tailings; this practice has caused destruction of the landscape.

Today it is possible to see several hectares devastated and destroyed due to the opening of pits, the construction of access roads, the building of infrastructures, the construction of tailings piles, all of which end up eliminating the landscape that once existed.

It was possible to identify the degradation of the landscape and alteration of the topography in order to create an area for the development of the activities carried out by the company. These activities drastically alter the landscape and topography and after the end of these activities it will be impossible to return to the same landscape and topographical features.

Destruction of the Flora

In the company, various quantities of solids are produced during the processing of mineral coal, and in some unit operations two types of product are obtained, namely tailings and concentrate.

As mining production increases, so does the production of solids in the processing plant, and the treatment given to tailings from mineral processing is to accumulate them in the form of conical piles.

In this process of forming piles and tailings, various species of trees are removed from a large area in order to make way for the tailings to be deposited. In this process, some of the plants are extremely rare and end up being lost because there is no replanting or reforestation process, putting certain species into extinction.

This practice of removing vegetation cover to make way for a tailings dump puts the ecosystem out of balance because fauna is extremely important for purifying the air, which is already polluted by dust.

It was possible to identify the removal of flora in order to create areas for the development of the activities carried out by the Jindal company. The reforestation process at this company has not been efficient because the species in instinct are not removed in such a way as to preserve them.

Noise pollution

The operation of the plant and the circulation of large machines end up generating vibrations and noises that pollute the typical sound of the region. These noises generated by men and their activities disturb the animals that are in their natural habitat, so the animals end up being driven away by the noise pollution and moving to other places far away from the epicentre of noise pollution.

Proposals for reducing the environmental impacts of coal processing and tailings disposal:

With a view to reducing the environmental impacts generated by the processing of mineral coal and the disposal of tailings at the Jindal company, the following proposals are made:

Air pollution:

- ✓ During the comminution process in the ROM, it is proposed that water be sprayed during the discharge of the material into the toothed roller crusher silo;
- ✓ The installation of air purifying inhalation devices along the area under study at critical points should be carried out.
- ✓ A process for recovering the coal inserted in the tailings is adopted as a way of reducing the calorific value of the tailings, thus avoiding spontaneous combustion of the tailings.

Soil and groundwater contamination:

- ✓ It is proposed that leaks in pipework and equipment be eliminated in order to reduce the amount of water lost and deposited in the ground;
- ✓ Adopt a system for removing stagnant water from the ground below the processing plant;
- ✓ Reduce the pressure of the feed to the spirals in order to reduce water drop and loss of solids;
- ✓ Propose treatment of water from boreholes consumed by the population in order to purify the water.

Destruction of the landscape and flora

- ✓ With regard to the destruction of the landscape in order to create areas for the disposal of tailings and the construction of infrastructures, it is proposed to reconstruct the environment in the same area and replant the species in instinct, with a view to restoring the vegetation to the devastated area.
- ✓ Until now, Jindal has not had a system in place for replanting trees, and the company proposes to create a system for replanting trees in areas that have already been mined.

- ✓ It is proposed that the tailings be returned to the open pits or craters in order to give the environment a suitable topographical appearance.

Final considerations

The interviewees realised that the environmental impacts generated by the mining companies on the population's routine are pronounced on windy days, when it is practically impossible to stay outside or even inside the houses because of the coal dust expelled from the mines. Houses, plants and animals are covered in coal dust, impacting on crop productivity and disrupting local cultural dynamics, since it is impossible to put flour and clothes out to dry

It was observed that a large number (47%) of the respondents perceive that living in Moatize is unsafe due to the harm that coal mining can bring to them. Dust was a frequently mentioned element among the respondents. It is often referred to as a popular form of coal dust, which causes health problems for those living in the municipality of Moatize, especially in the neighbourhoods close to the mine.

In general, respondents don't seem to have a perception of the impact of coal dust on water quality, since their assessment is linked to access to and supply of water and not to water quality per se.

The perception of the environmental impact and the magnitude of the health risks among residents of Moatize affected by opencast coal mining does not seem to be as heterogeneous as might be expected due to the influence of the socio-economic characteristics of the interviewees. The analysis of the data showed that most of the residents who were surveyed do not seem to realise that environmental exposure to coal dust occurs practically everywhere in the municipality of Moatize because of the dispersion of the coal particles, although the mines themselves are the biggest threat, because of the explosions and the accumulation of the "piles" of coal waste generated during the mining process and which are deposited in this physical space.

Throughout this study we have tried to introduce a communicative element of reflection and awareness about the impact of open-cast coal mining to a small section of the population environmentally exposed to coal dust. In order to promote transformations in the individual and collective attitudes of risk groups, risk perception studies often serve to subsidise risk communication projects and strategies. However, it is imperative to investigate whether education and information can overcome residents' misconceptions about the risks they face from exposure, as even well-structured information measures may not be able to change some people's risk behaviours.

At the same time, the results of the study show that it is imperative that there is an increase in the institutional responsibility of companies and government bodies, providing health care and promoting environmental interventions for the population of Moatize, directly affected by the precarious living and health conditions and coal dust, as well as a more in-depth re-discussion of the political and ethical aspects linked to the resettlement of the most exposed population groups.

Suggestions

On the basis of the above, we believe it is pertinent to make a few comments and propose the following recommendations:

i. Conduct sustained research: Research should be conducted in a sustained manner to assess and mitigate all negative environmental impacts associated with coal mining and exposure to coal dust.

ii. Institutionalising the notification of exposure to coal dust: There is a need to study different ways of dealing with the different social groups living in the regions affected by coal mining in order to carry out prevention and awareness campaigns on the potential impacts of coal mining on the environment and human health.

iii. Applying the concept of ecosystem services in EIAs: Recognising the importance of incorporating the concept of ecosystem services into

Environmental Impact Assessments.

iv. Promoting the institutionalisation of complaint and conflict management mechanisms between the affected communities and the mining companies: The complaints made in general by the communities in the areas covered by the study show a lack or ignorance of the existence of institutionalised mechanisms for dialogue between the communities in the area and the mining companies.

Bibliographical references

Araujo, L. C. G. de (2009); *People Management - Strategies and Organisational Integration.* 2 ed. São Paulo: Atlas.

Abrão, P. C., Oliveira, S. L. (1998). *Engineering Geology:* Mining (2ª . ed.). São Paulo, Brazil: ABGE - Brazilian Association of Engineering Geology, p. 431- 438.

Bardin, A. J. (2011) *Metodologia de Trabalho Científico* (21ª . Ed.). São Paulo, Brazil: Cortez Editora, p.15.

Bellia, V. (1996). *Introduction to environmental economics.* Brasilia, Brazil: IBAMA.

Calaes, G. D. (2006). *Strategic Planning. Competitiveness and Sustainability in the Mineral Industry.* Rio de Janeiro, Brazil: CETEM/MCT/CNP, p. 242.

Coelho, M. C. (2004). Environmental Impacts in Urban Areas. In A. J. Guerra, S. B. Cunha. *Impactos Ambientais Urbanos no Brasil.* (2ª . ed.). Rio de Janeiro, Brazil: Bertrand, p.19-45.

Gil, A. C. (2007). *Como elaborar projetos de pesquisa.* (4ª . ed.). São Paulo, Brazil: Atlas.

Gil, A. C. (2019). *Como elaborar projetos de pesquisa.* (6ª . ed.). São Paulo, Brazil: Atlas.

Girodo, A. C. (2005). *Mining.* (4ª . ed.). Porto, Portugal: Decada das palavras, p. 168.

Goldemberg, J. (1998). *Energy and development.* São Paulo, Brazil: Atlas.

Lakatos, E. M. (2007). *Sociology of Administration.* São Paulo, Brazil: Atlas.

Lakatos, E. M. & Marconi, M. A. (1991). *Metodologia Cientifica.* (2ª . ed.). São Paulo, Brazil: Atlas, p. 242.

Marques, J. R. (2005). *Urban Environment.* Rio de Janeiro, Brazil: Bertrand.

Mendonça, F. A. (2004). *Geography and the Environment* (7ª . ed.). São Paulo, Brazil: Contexto.

Pondja Jr, E. A., Persson, K. M. & Matsinhe, N. P. (2017). *Assessment of coal mine water in Moatize by static and leaching tests Sustain.* Maputo, Mozambique: Palavras

Prodanov, C. C., Freitas, E. C. (2013). *Research methods and techniques for academic work.* (2ª . ed.). Madrid, Spain: Pearson Hall.

Redaelli, L. L., Cerello, L. (1998). *Engineering Geology:* Excavations. (2ª . ed.). São Paulo, Brazil: ABGE - Brazilian Association of Engineering

Geology, p 311-330.

Schumann, W. (2008). *Guide to Minerals.* São Paulo, Brazil: DISAL, p. 127.

Severino, A. J. (2010). *Metodologia de Trabalho Científico.* (21ª . ed.). São Paulo, Brazil: Cortez Editora, p.46.

Valle, C. E. (2004) *Qualidade Ambiental.* (5ª . ed.). São Paulo, Brazil: SENAC.

Vilelas, J. (2009). *Research: The Process of Building Knowledge.* Lisbon, Portugal: Silabo.

Wicander, R., Monroe, J. S. (2009). *Fundamentals of Geology.* São Paulo, Brazil: Cengage Learning, p. 508.

MIX
Papier aus verantwortungsvollen Quellen
Paper from responsible sources
FSC® C105338

Printed by Books on Demand GmbH, Norderstedt / Germany